Growing Windowsill
Orchids

PHILIP SEATON

Contents

Introduction

The mystery and romance of orchids make them hard to resist. With around 25,000 species, orchids are possibly the largest plant family in existence: around one in ten flowering plants is an orchid. They display an amazing variety of flower types ranging from the flamboyant 'chocolate box' *Cattleya* to the frankly bizarre-looking slipper orchids.

Orchids were once the exclusive preserve of the rich, who employed their own growers to cultivate them in large hot houses. However, people have recently discovered that, especially where they have central heating, many can be grown just as easily as other houseplants. Specialist orchid nurseries, garden centres and supermarkets offer an ever-increasing range of affordable exotic blooms for your windowsill. Although they have a lingering reputation as being difficult to grow, in reality orchids are more robust than many other popular houseplants. If they are provided with the right conditions, they will grow and bloom, and give enjoyment for many years.

Be warned though: orchids are addictive! So thank you to my children Richard, Anna, Claire and Joseph, for putting up with their father's orchid obsession without complaint.

Philip Seaton

Buying your orchid

When buying an orchid there are some key things you should look for to ensure that you choose a plant that will stay in bloom for as long as possible. Choose a healthy-looking plant that has some open flowers, so that you know exactly what you are buying, and reject plants with deformed or bruised flowers.

Look for plants with some unopened buds at the tip of the flower spike; not only does this mean that there are more flowers to come, it probably also indicates that the plant has not been on the shelf, possibly in dry conditions, for too long. If you search carefully you may find additional flower spikes beginning to develop. When choosing a slipper orchid that only carries a single flower, you should aim to buy a plant with a flower that has recently opened. You are looking for a plant that has recently arrived at the point of sale. Yellowing buds, and indications that some buds have dropped off, are not good signs.

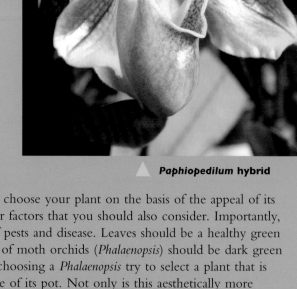

▲ *Paphiopedilum* **hybrid**

◄ **Scarlet Cambria**

Naturally you will wish to choose your plant on the basis of the appeal of its flowers, but there are other factors that you should also consider. Importantly, the plant should be free of pests and disease. Leaves should be a healthy green and not yellowing. Leaves of moth orchids (*Phalaenopsis*) should be dark green and relatively stiff. When choosing a *Phalaenopsis* try to select a plant that is sitting upright in the centre of its pot. Not only is this aesthetically more pleasing, it can make it easier to re-pot your plant at a future date.

You also want a plant with a healthy root system. This can be more difficult to judge without tipping the plant out of its pot to have a look – something the seller might not appreciate. Beware of plants that have been over-watered, with compost that has become a wet and soggy mass: this can lead to the death of the roots and the eventual demise of the whole plant. *Phalaenopsis* are frequently sold in clear pots making it easy to see the compost and roots within the pot. Their roots also tend to enjoy growing out of the pot and into the air. In both instances you are looking for plump silvery roots with green or reddish-green tips.

If your plant is of the type that has fat green bulbs (technically they are called pseudobulbs), look for shrivelling. A slight shrivelling of the bulb may indicate that the plant requires watering, in which case a light spraying of the compost with water will revive the plant and make the bulb swell and plump out. Heavily wrinkled bulbs can indicate either severe under- or over-watering and should be rejected in favour of a plant that doesn't show these symptoms.

If you have the opportunity, why not visit an orchid show (there are many local orchid societies) or a specialist orchid nursery? You may have to pay a little more for your plants but there is likely to be a bigger selection, the plants are often larger, and you can get free advice from the nursery staff. If you want to learn more about orchids, it is worth considering joining your local orchid society.

There are bargains to be had from time to time, from market traders for example, who may offer good quality plants at very low prices. Remember, however, that *Phalaenopsis* in particular do not appreciate being stood outside in the cold.

Getting your plant home

During the winter months you should be aware that a sudden fall in temperature could lead to bud drop. Ideally, flower spikes should be carefully wrapped in tissue paper to protect them while transporting your plant home, although they are often sold in clear plastic sleeves that also offer some protection against accidental damage to the flowers.

Once you get your plant home

Carefully remove the tissue paper to avoid damaging the flowers and flower buds. If your plant is in a plastic sleeve, carefully cut the sleeve away with a pair of scissors rather than risk damaging the flowers when extracting your plant from the sleeve.

Author's tip

Orchids are surprisingly tough plants and are slow to react to neglect. Many retailers guarantee their plants for a certain period, so if your plant dies within a short time after being purchased, despite it having been given the recommended basic care, you may be able to return it to the seller.

Cerise *Dendrobium phalaenopsis*
▼

Caring for
your orchid

General care advice on how to look after your orchid is often supplied on the pot label, in terms of temperature, lighting and how often to feed your plant. However, more detailed information, is often lacking.

In many respects your home offers an ideal environment for growing orchids. It is usually a consistently warm temperature, and seldom, if ever, gets too hot or too cold. It also offers a number of different micro-climates in which your plants can grow.

Light

All green plants need sufficient light for photosynthesis – but not so much that it burns their leaves. Most orchids will not tolerate the bright sunlight of a south-facing window during the summer months. Many orchid species originate from the rainforest where they are adapted to growing in the shade beneath the tree canopy, so do not thrive in direct light. However, unlike some other houseplants, they generally do not like the low light conditions away from a window. Many appreciate the early-morning or late-evening light of an east or west-facing window. North-facing windowsills are coolest, and can sometimes provide sufficient light for some orchids during the summer months.

When you first buy your orchid it is worth experimenting a little to find the best location. What is ideal for a *Phalaenopsis*, for example, may not be suitable for a *Miltonia*. Plants can be moved from one room to another. They can be moved to a sunnier windowsill with more available light for the winter months and moved back to a shadier place in the summer.

As new flower spikes grow and begin to produce a straight or gracefully arching stem it is a good idea to maintain your plant in the same orientation regarding the light. This can also help avoid bud drop. Train the stem up a plant-support, gently tying it to the support at intervals.

Miltonia shaded by curtain

Temperature

Some orchids prefer warm conditions while others thrive in cooler environments and may even appreciate a spell outside in the summer. With the advantage of central heating, orchids that prefer warm conditions, such as *Phalaenopsis*, can easily be grown in the home on the windowsill. Avoid cold draughts and bring them into the room from behind the curtains on cold nights.

More specific advice regarding suitable temperatures is provided in later sections where the different types of orchid are described.

Humidity

Most orchids that are offered for sale as houseplants have their origins in the humid tropics and grow best in a moist environment. The atmosphere in the home, especially in the winter when the central heating is on, may be rather dry and can lead to bud drop. This can easily be resolved by standing the plants on a tray of moist gravel, or growing them in groups, perhaps with other houseplants, to raise the humidity. Most orchids will also enjoy the warm, humid, bright conditions found in bathrooms.

Author's tips

Once you have caught the orchid-growing bug you may be tempted to grow some of the smaller 'botanicals' offered for sale at orchid shows. Often these require the more humid environment provided by a terrarium that can be constructed from an aquarium tank or a large glass bottle.

Paphiopedilum are not the only orchids that can have attractive leaves. 'Jewel orchids' are grown specifically for their beautifully patterned foliage rather than their relatively insignificant flowers. Members of the genus *Macodes* and *Ludisia* are frequently offered for sale as jewel orchids. They are usually grown in a moisture-retentive, peat-based compost. The compost should be kept consistently moist but not wet.

Orchids on a tray of moist gravel to raise humidity

Leaves of *Macodes petola* ▶

8

Watering

Maintaining a strong healthy root system is the key to success with orchids. Healthy roots lead to a healthy plant, which in turn leads to a good display of flowers. Instead of having the familiar fibrous structures, coated with root hairs, of plants growing in the soil, epiphytic orchids (*see* page 54) have a special water-absorbing tissue called the velamen designed to rapidly soak up moisture like blotting paper as soon as it becomes available. This silvery velamen will turn green on watering and, if you look closely, you will see small elliptical areas that remain white: these pore-like areas are called lenticels. Roots need to breathe too, and lenticels make it possible for oxygen and carbon dioxide to diffuse in and out of the roots. Any compost needs to contain sufficient air spaces to enable roots to obtain oxygen and carbon dioxide.

Knowing when, and when not, to water is probably the most common cause of concern for novice and more experienced growers alike. Because watering should depend on environmental conditions (if the weather is warm or cool for example, or if it is summer or winter) and the type of compost, there can be no hard and fast rules for frequency of watering. However, there are some basic rules that can be followed.

In common with other houseplants, always bear in mind that the most frequent cause of death in orchids is probably over-watering. The golden rule is: **if in doubt, don't**. Although not to be viewed as a recommendation, with a few exceptions, orchids can often survive an amazing amount of neglect. *Phalaenopsis* in particular are surprisingly tough, and can survive for long periods without water. A thorough watering once a week in warm summer conditions, combined with a regular light spraying of the leaves and any aerial roots, can be used as a general guideline. Always remember to remove excess water from the crowns of *Phalaenopsis* (use a tissue to soak up any excess) and refrain from spraying the plant when it is in direct sunlight to avoid burning the leaves.

Orchids benefit from a regular light spraying of the surface of the compost. If your orchid has pseudobulbs (as illustrated) spraying is often sufficient to prevent them shrivelling during the winter months.

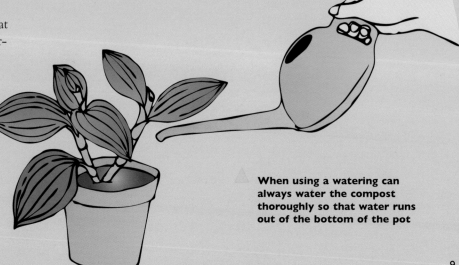

When using a watering can always water the compost thoroughly so that water runs out of the bottom of the pot

Use rainwater to water your orchid when possible, especially if you live in a hard-water region. The water should be left over night to reach room temperature. If you are using tap water, leaving the water in a container for a few hours also has the advantage of allowing time for any residual chlorine to evaporate.

How do you know if a plant needs watering? Does the compost look dry? Remember that even though the surface is dry it may be much wetter lower down in the compost. Pick up the pot – if it feels light the plant probably needs watering. Water the plant thoroughly and pick up the pot again. It should now feel heavier than before. It is worth noting that even when thoroughly wet, orchid compost is much lighter than the compost used for your other houseplants.

Symptoms of under and over-watering include shrivelling of the pseudobulbs, or wrinkled leaves in *Phalaenopsis*. We tend to assume shrivelled pseudobulbs are due to a lack of water, and this is often correct. However, over-watering will lead to the death of the root system, and the plant will similarly be starved of sufficient moisture. Many orchids are sold in clear plastic pots. One advantage of these pots is that you can see what is happening beneath the surface of the compost and gauge whether the roots are healthy.

One of the benefits of growing orchids in the home (rather than in a greenhouse for example) is that they are under constant, often daily, observation and any problems can be spotted early and prompt action taken. Never be afraid to tip a plant out of its pot to see what is happening to the root system and to examine the state of the compost. This will help you judge whether your plant is suffering from under or over-watering. It has been suggested that because roots can be photosynthetic the plants will benefit from the extra light the roots obtain in clear plastic pots. They do however look much more attractive in a container of your choice.

The compost should allow water to drain through rapidly. You can either water using a small watering can, or, where the compost has been allowed to become very dry, you may need to stand the pot in a bowl of water for half an hour to allow the compost to take up the moisture.

Feeding

Your orchids should only be fed whilst actively growing. This is easy to determine with orchids such as *Phalaenopsis*, whose roots tips turn green or reddish-green when growing. With other types you should examine the roots carefully in the compost (where there is a good root ball, don't be afraid to tip the plant out of its pot to examine the roots from time to time).

There is an ever increasing range of fertilisers sold as powders, liquids and sprays aimed specifically at the orchid growing market

With a few exceptions, because they grow in the forest canopy where nutrients are limited, orchids are not normally heavy feeders. Normal house plant or garden fertiliser is generally suitable and should be used at one quarter strength. You can also buy specialist orchid fertilisers, in which case you should follow the instructions on the package/bottle. To avoid a build up of salts in the compost, water with tap water (or rainwater if available) every third watering. At the same time as they are absorbing water through their roots, orchids take up a wide range of other essential nutrients. Chief amongst these are nitrogen (chemical symbol N), phosphorus (chemical symbol P) and potassium or potash (chemical symbol K), which are required by the plants in relatively large amounts. All fertilisers have information about the amounts of the different nutrients printed on the outside of the bottle or packet in the form of the NPK ratio. A typical all-round fertiliser might have an NPK ratio of 20:20:20.

Nitrogen promotes the production of lush, soft, dark green leaves and may be used to stimulate vegetative growth early in the year. Because the fir bark used in many orchid composts lowers the available nitrogen, some growers recommend using a fertiliser with an NPK ratio of 30:10:10 at the beginning of the growing period.

However, lush green growth isn't necessarily associated with flowering. You also want a strong and disease-resistant plant. Phosphorus promotes the formation of a healthy root system and flowering, and potassium promotes strong growth and flower development. Thus, to stimulate flowering, it is recommended that you use a fertiliser containing proportionally less nitrogen and more phosphorus and potassium later in the year, with an NPK ratio of 10:30:20 for example. Tomato fertilisers are similarly high in phosphorus and potassium and are ideal for stimulating flower production in *Cymbidium*.

Composts and re-potting

Your orchid will have been grown in one of any number of different composts based on peat, bark and charcoal, and, possibly less frequently today, rock wool. In addition you may discover perlite, sphagnum moss or pieces of sponge. Whatever

A range of specialist orchid composts is now widely available at reasonable prices

the mixture, the aim is to produce a very open compost that retains moisture at the same time. All plants need air around their roots to allow them to breathe, but orchids in general, and epiphytic orchids in particular, need more air than usual. Always use orchid compost; general composts for growing houseplants are not suitable. Orchid composts can be purchased in small amounts from specialist orchid nurseries and most garden centres or you can make up your own compost.

When to re-pot

Depending on the type of orchid, re-potting is generally best carried out in the spring when the roots are active. It is a mistake to assume that a plant always requires a larger pot each time; a smaller container is sometimes more appropriate. As a general rule, over-potting is a mistake. When you remove your plant from its container you will often find that the roots have curled around the pot's edge and have not penetrated to the centre of the compost. This is because they prefer the dryer conditions around the outside of the pot. When a plant has not been re-potted for some time, the compost in the middle of the pot can begin to break down, and, if you are not careful, it will turn into a constantly wet and soggy mass.

When planning to re-pot, think ahead. If the plant is dry, water it the day before, as it will probably take some time to re-adjust and begin to take up moisture. Soak the new compost overnight too. Having selected the correct size of container, place a small amount of drainage material (such as expanded polystyrene chips) in the bottom. Pour in a small amount of compost to cover the drainage material and place the plant in the desired position. Pour in the compost, gently easing it in around the roots, repeatedly tapping the pot gently on the bench so that the compost settles. Fill to within two or three centimetres (an inch) of the pot rim. Don't compact the compost with your fingers. The plant should sit firmly in the pot; if it is loose, the plant should be staked to stop it wobbling.

Author's tips

Making your own compost

Making your own compost is easy to do, but it is less easy to explain what the compost should look and feel like, as it often depends on the quality of the components, and the bark in particular. As a rough guide a general purpose orchid compost can be made from three parts bark, one part sphagnum moss, one part perlite, and one part charcoal.

1. **Pumice: can be added to the compost for plants with a vigorous root system such as *Cymbidium*.**

2+3. **Pine bark: sieve the bark before use to get rid of any dust – this tends to settle to the base of the pot and block the drainage holes. Large-grade bark should be used for plants with thick roots such as *Cymbidium*, and a medium-grade for plants with finer roots such as *Odontoglossum*. General purpose bark sold in garden centres for paths and as a soil conditioner is not suitable.**

4. **Seramis® clay granules: can be incorporated to maintain an open compost for finer rooted orchids.**

5. ***Sphagnum* moss: absorbs and retains a lot of moisture. It is usually purchased dry in blocks, which expand considerably when water is added. The Sphagnum moss will usually have been imported from New Zealand or Chile.**

6. **Charcoal: keeps the compost 'sweet' by adsorbing harmful breakdown products.**

7. **Perlite (coarse grade): retains moisture while retaining an open texture. You should always wet perlite before use as inhaling the dust may be harmful.**

Safety note

Always wear a suitable mask when mixing dry, dusty components.

Phalaenopsis –
moth orchids

A large well-grown plant can produce a breathtaking display of elegant blooms that last, in perfection, for up to three months or more. They may flower two or even three times a year. Due to the efforts of hybridisers around the world, moth orchids are available in a dazzling array of markings and colours including whites, pinks, yellows and clarets. *Phalaenopsis* are closely related to *Doritis* with which they are frequently hybridised to produce the genus *Doritaenopsis,* which typically has dark cerise blooms on upright spikes. Recently, crosses with the diminutive flowers of *Phalaenopsis equestris* have led to the production of a range of charming miniature moth orchids.

New growers are often puzzled by the unruly silver roots that refuse to confine themselves to the pot, and, illustrating their epiphyte heritage, worm their way into the atmosphere. They often ask if the roots should be pushed down into the compost. The answer is definitely 'no'. Aerial roots should be left to grow where they will. They also act as a good indication of whether the plant is in growth or not: when the plant is actively growing the roots have green or reddish-green tips.

Watering

As a general guide, *Phalaenopsis* should be watered regularly once or twice every week during the growing period from spring through to autumn. The compost should be allowed to dry out in between watering but not to the extent it becomes bone-dry. Either pour water through the compost or immerse the pot in a bowl of water (which is at room temperature) and leave for a few minutes before removing the pot and allowing the compost to drain. The roots and leaves also benefit from a daily spray with tepid water at this time of year. Care should be taken not to allow water to remain in the crown of the plant, as this can rapidly lead to the rotting of the crown. Any water that does remain should be removed promptly with absorbent tissue. Leaving water on the leaves in bright sunlight may also cause burning, as the water acts as a lens focusing the heat on the leaf. A slight wrinkling of the leaves may indicate that the plant needs more water. Watering should be reduced in the winter to once a week or possibly less.

Feeding

Feed your orchid every other watering during the growing period i.e. when the roots have active green or red-green tips. For more details about fertilisers see page 11.

Temperature

For the best results a minimum temperature of 18°C (64°F) and a maximum of 26°C (80°F) are often recommended. *Phalaenopsis* will, however, tolerate lower temperatures (central heating is, for example, not essential) common in the home at night in the winter months. Plants can be stood on a sunny windowsill in the winter, and brought inside the room when the curtains are drawn in the evening.

Light

Grow in a bright location, out of direct sunlight. A south-facing location should be avoided in the summer months, unless the plants are shaded by a net curtain for example.

Humidity

Phalaenopsis enjoy warm humid conditions. In a dry centrally-heated room the plants will benefit from standing their pots on trays of damp gravel.

Miniature orchids, like this *Phalaenopsis*, can be a real bargain!

After flowering

Once your *Phalaenopsis* has finished flowering you have two options. If you look closely at the flower spike you will see that it is divided into sections. At the beginning (and end) of each section is a small scale with a bud beneath: this is called a node. You can either cut the flower spike close to its base, or just above a node, which may induce the growth of a new flowering branch. If the plant is growing strongly, with large rigid green leaves, cut the flower spike off above the third node. The plant may then, in time, produce a strong branch and many more flowers. However, if the plant has been flowering for some months, it may be better to cut the spike close to the base, and wait for a new flower spike to appear.

If you have bought a plant in flower and have cut the flower spikes off at their base when they have finished flowering, you can normally expect to see new flower spikes emerging in September or October, stimulated by cooler temperatures.

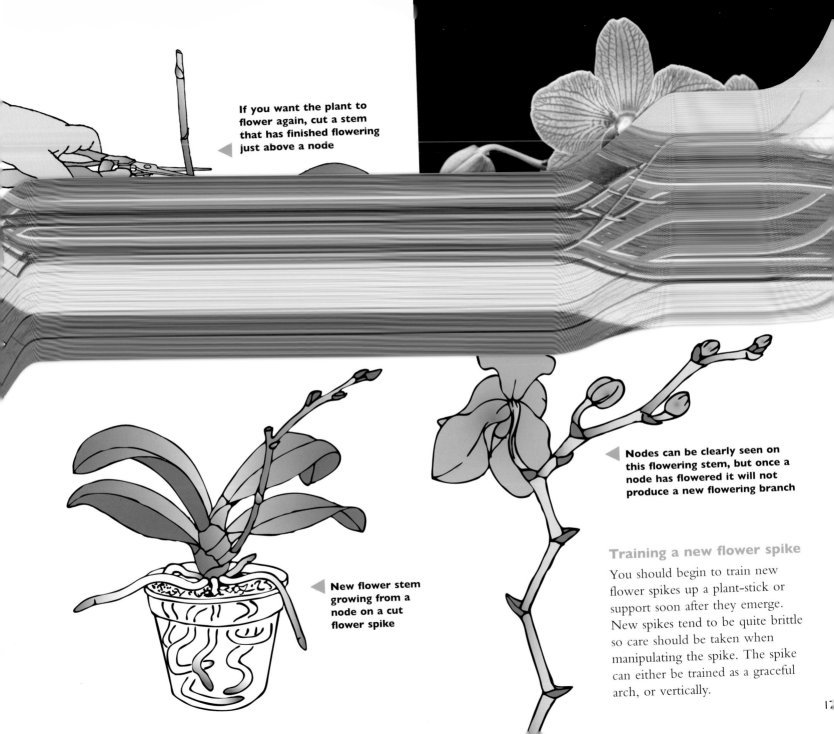

If you want the plant to flower again, cut a stem that has finished flowering just above a node

Nodes can be clearly seen on this flowering stem, but once a node has flowered it will not produce a new flowering branch

New flower stem growing from a node on a cut flower spike

Training a new flower spike

You should begin to train new flower spikes up a plant-stick or support soon after they emerge. New spikes tend to be quite brittle so care should be taken when manipulating the spike. The spike can either be trained as a graceful arch, or vertically.

17

Occasionally, instead of a new branch developing from a node on a flower spike, a new plant will develop in its place. Such plants are called keikis (keiki is the Hawaiian word for baby, pronounced 'kay-key'). This miniature plant can be allowed to grow and once it has developed two or three leaves and a healthy root system, it can be carefully cut away from the parent together with a small piece of the flower spike. It can then be potted in orchid compost and grown on to flowering size in one or two years. Keikis are clones of the parent plant and, as such, will produce identical flowers to their parent.

Keikis can be artificially induced using keiki paste, which contains hormones called cytokinins that induce plantlet formation. Hormone rooting powders are not suitable, as they contain auxins, a different group of plant hormones that induce the initiation of roots.

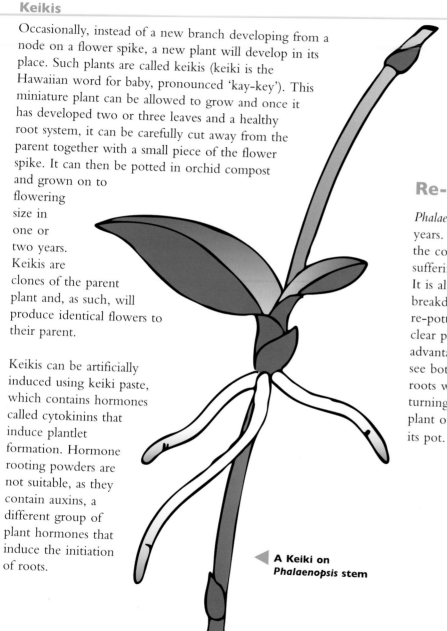

▲ A Keiki on *Phalaenopsis* stem

Author's tip

Orchid rescue. If your *Phalaenopsis* does begin to look sick, don't despair, you can often rescue the plant by re-potting into fresh compost, and then spraying the roots and foliage daily until new roots, and sometimes a new shoot, appear.

Re-potting your *Phalaenopsis*

Phalaenopsis should be re-potted regularly – at least every two years. The best time to re-pot is normally in the spring but if the compost is obviously breaking down and the plant is suffering, re-potting can be carried out at any time of the year. It is always a mistake to wait until the compost begins to breakdown and problems begin to occur before re-potting. *Phalaenopsis* are often sold in clear plastic pots and these have the advantage of allowing the grower to see both the compost and the roots without turning the plant out of its pot.

1 **Carefully remove plant from its pot**

- If your plant is growing in a plastic pot it may be possible to gently squeeze and deform the pot to loosen the compost.

- While gently holding the plant, invert the pot over a tray to catch the old compost.

- Gently remove the compost from around the roots.

- Trim any old roots with a sterile sharp knife or scissors, and cut off any old flower spikes close to the base.

- Select a suitable sized container (this could simply be the pot in which the plant has been growing, in which case it should first be cleaned thoroughly). Don't be tempted to use a larger container unless you cannot easily introduce the roots into the old pot. If the root system is poor, you may be better advised to use a smaller container.

- You may wish to introduce some drainage material into the bottom of the pot, but if it is the correct size of pot, this may not be necessary.

- Introduce the plant into the pot and pour in the moist compost around the roots, gently teasing it between the roots with your fingers, tapping the container gently at intervals to allow the compost to settle. Do not push the compost down too firmly: the roots need plenty of air circulation.

- Do not attempt to force any roots that were previously growing outside the pot into the container.

- Spray the surface of the compost and the roots daily until your plant is established.

- Once the plant is established, water regularly.

The plastic pot can be placed inside a more ornamental container if you wish.

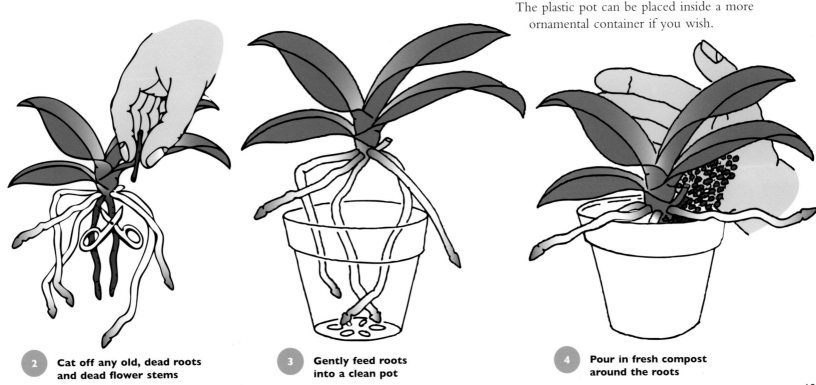

2 Cat off any old, dead roots and dead flower stems

3 Gently feed roots into a clean pot

4 Pour in fresh compost around the roots

Cymbidium

Cymbidium plants are easily distinguished from other orchids by their long, strap-like leaves. Their large long-lasting waxy blooms are familiar as the individual boxed flowers that make popular gifts on Valentine's or Mother's Day. These blooms are from the so-called 'standard' *Cymbidium* plants, which, due to their large size are generally more suitable for growing in a greenhouse or conservatory. Much more suitable for growing in the home are the 'miniature' *Cymbidium* plants. Their tall spikes of a dozen or more flowers can last six weeks, or even longer in a cool environment.

Temperature and flowering

Cymbidium prefer cool conditions and should be kept at a temperature of around 10°C (50°F) in winter, and a maximum of 24°C (75°F) in the summer.

The most commonly asked question about *Cymbidium* is, 'How can I get my plant to flower again?' Typically, people who grow these orchids at home produce vigorous plants with healthy, dark green leaves but no flowers.

Once your plant has finished flowering, cut the flower spike near to the base. To encourage the plant to flower again the following year, use a feed with an 'NPK ratio' that is high in phosphorus and potassium (such as a tomato fertiliser) throughout the summer and early autumn, making sure your plant gets a drop in night-time temperature in the autumn. *Cymbidium* are often better grown outside in summer in a shady spot, or in a shaded greenhouse (to prevent scorching of the leaves). Cool September nights will stimulate flowering, but it is important to remember to bring your plants inside before the first risk of frosts. Reduce watering to a minimum over the winter months, without allowing the compost to dry out completely.

Miniature *Cymbidium* can be grown on the windowsill. They tend to have more flowers than their standard *Cymbidium* cousins. ▶

◀ **A lovely green standard *Cymbidium* with large blooms and full rounded petals. Standard *Cymbidium* make ideal subjects for growing in the conservatory.**

Problems

- *Cymbidium* are particularly susceptible to red spider mites (*see* page 48), which are usually found on the under-surface of the leaves. When you turn the leaf over, infestations appear as a silvery mottled sheen, sometimes turning brownish in colour.

- If you grow your *Cymbidium* outside in the summer months remember to check regularly for slugs and snails.

- Brown tips on the leaves can be caused by allowing your plants to dry out too much between waterings. Trim the ends of the affected leaves to a v-shape with clean scissors.

- Sunken brown markings on the leaves in a diamond pattern may indicate that the plant is suffering from a viral infection.

Emerging flower spikes are bullet-shaped, rather than the pointed shoots typical of new leaf growth. Once the new spikes appear, you should begin to train them up a plant-stick or support, preferably using soft string or raffia to tie them in. Insert the support carefully away from the edge of the pot (thereby avoiding the new roots, which tend to coil around the edge of the pot). The young flower spikes of *Cymbidium* are surprisingly brittle and easily broken off so do take particular care when training them up supports. When staking your flower spikes always place a plastic cap on the top of the support to prevent accidental damage to your eyes when bending over the plant.

Old flower spikes should be cut off close to the base

Cymbidium display at a show in South Africa

Re-potting your *Cymbidium*

Unlike other orchids, as long as the compost isn't breaking down, *Cymbidium* plants can remain in the same pot for two or three years, and don't require annual re-potting. Indeed, they often appear to enjoy being pot-bound. When required however, plants should be re-potted in the spring when the roots start to become active. Re-potting a large *Cymbidium* can be a daunting task, especially if delayed until it has become severely pot-bound. If it is an old plant grown in a clay pot and the roots are adhering firmly to the pot walls, it may be necessary to break the pot with a hammer. If it is growing in a plastic pot, you might have to slit the walls of the pot with a sharp knife (with care!) to be able to remove the plant.

Black and red boxed orchid for sale in Mexico. Although native to South-East Asia and China, *Cymbidium* are often seen for sale in the markets of Mexico City as cut flowers. Here the flowers are often dyed with artificial colours.

Orchids have been cultivated in China and Japan for many centuries, a practice that remains popular today. They are grown in beautiful ceramic pots to enhance the appreciation of their elegant foliage. *Cymbidium goeringii* is grown for the delicate perfume of its spring flowers.

This *Cymbidium* has been repotted in a mixture of coarse bark, pumice and charcoal

Author's tip

Old pseudobulbs can be placed in fresh compost, and will often shoot, and can eventually be grown on into mature plants.

Often you will be confronted with an impenetrable root ball. Using a clean sharp knife, a large plant can be divided into two or more smaller ones. However, it is vital that each individual plant should retain at least three or four good size backbulbs (leafless pseudobulbs). Gently tease the roots apart with your fingers. Dead roots should be cut away using clean sharp scissors or secateurs. Live roots are generally white and if actively growing they will have rounded cream-coloured tips. Dead roots are generally brownish and dry with the outer layer (the velamen) often separating easily from the tougher fibrous core of the root. Roots that are too long to fit into the new pot can also be trimmed.

Re-potting also provides an opportunity to remove the dead leaf bracts surrounding the old pseudobulbs, although this can be done at any time. Bracts can be split from top to bottom and carefully peeled away. Take care not to damage any underlying shoots.

Select a pot of a size that allows sufficient space for one or two years' new growth. Place the rearmost bulb against the edge of the pot. The bulbs should sit on the surface of the compost and not be buried. Gently fill the pot with compost around the roots, making sure not to compact it. Do not feed for the first three months, or until new roots are established.

Author's tip

Plant diseases can be spread using dirty pots. Always use either new pots or pots that have been thoroughly cleaned.

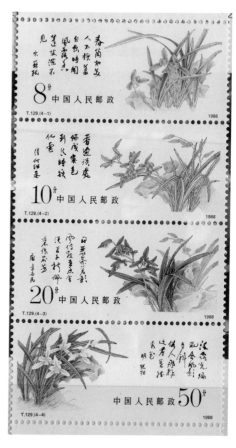

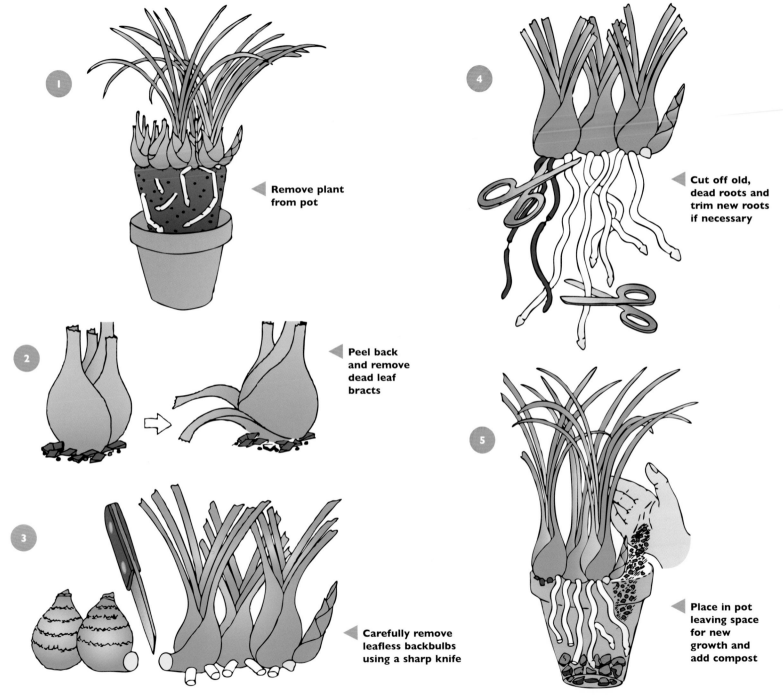

1 Remove plant from pot

2 Peel back and remove dead leaf bracts

3 Carefully remove leafless backbulbs using a sharp knife

4 Cut off old, dead roots and trim new roots if necessary

5 Place in pot leaving space for new growth and add compost

25

Cambria orchids

Orchids can be very promiscuous, not only readily forming hybrids between different species within the same genus, but also between genera. Cambrias are a large and rapidly expanding group of easy to grow orchid hybrids, incorporating a confusing array of many different but related orchid species from the orchid genera *Odontoglossum, Oncidium, Cochlioda, Miltonia* and *Brassia*. You may come across such exotic names as *Beallara, Burrageara, Colmanara, Odontonia, Vuylstekeara* and *Wilsonara* for example. What they all share is that they have *Odontoglossum* somewhere in their parentage.

Cambrias are often found in a wide range of dazzling colours, often with beautiful patterns and markings reflecting their *Miltonia* heritage. Plants with *Cochlioda* in their 'blood' have flowers in beautiful shades of red from a deep burgundy to scarlet, whereas crosses containing *Oncidium* genes produce golden yellows. Hybrids containing *Brassia* in their parentage produce attractive spidery flowers.

All Cambrias have plump, egg-shaped pseudobulbs. When buying these types of orchid ensure that the pseudobulbs are not too wrinkled and certainly not shrivelled.

Odontoglossum

Odontoglossum or butterfly orchids are one of the most important parents of many Cambrias. Although the lovely crystalline white flowers of *Odontoglossum crispum* itself are seldom seen in cultivation, it has been an important parent of many other white-flowered varieties. *Odontoglossum* prefer cool conditions, 10 to 20°C (50 to 68°F), as they are from the high altitudes of the misty mountains of Central and South America. Hybridisation with other genera has produced plants that will thrive in warmer conditions.

Miltonia

Miltonia (strictly speaking they are called *Miltoniopsis*) are commonly known as pansy orchids because of their attractive faces and patterning. They typically have long, thin greenish-grey leaves and their pseudobulbs are more flattened than those of the Cambria orchids to which they are related. They are slightly more challenging to grow than the other orchids in this group and will soon rot if over-watered.

Oncidium

Oncidium or 'dancing ladies' typically produce long branching sprays of small bright-yellow flowers that have enormous lips in proportion to the size of the flowers. In their native Latin America they are referred to as *lluvia de oro* (golden rain). The branching habit of *Oncidium* species has been used to produce many multi-flowered hybrids in a rainbow of colours.

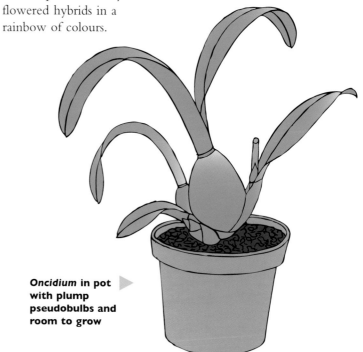

Oncidium in pot with plump pseudobulbs and room to grow

Watering

The golden rule for watering is: **if in doubt – don't!** Generally speaking, watering once a week is sufficient, although in warm conditions plants will dry out more quickly and should be watered more frequently. During the summer months, when the plants are actively growing, they can be watered by standing the pot in a bowl of luke-warm (preferably rain) water for a few minutes. In winter they have a distinct resting period and require less water. If the pseudobulbs begin to shrivel at this time of year spraying the surface of the compost is usually sufficient to plump them up again.

In this group of orchids the new leaves and shoots can become folded in a concertina pattern as they emerge. This is caused by infrequent watering, where the compost is allowed to become excessively dry between each watering. Although the concertina pattern will persist as the leaves age, keeping the compost more consistently moist (not wet) will prevent this happening in any new leaves.

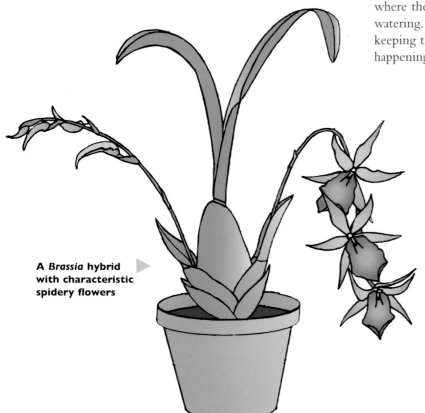

A *Brassia* hybrid with characteristic spidery flowers

▲ **Two different colour varieties of Cambrias**

Oncidium are often called 'dancing ladies'

Re-potting

If a plant is well cared for, each new pseudobulb will be larger than that of the previous year, and certainly not smaller. All of the Cambria types can be re-potted in the spring as new growth and roots begin to emerge. Sufficient space should be allowed in the pot for at least one year's growth. With their finer roots, Cambria types should be potted in a compost containing a much smaller grade of bark than *Cymbidium* or *Phalaenopsis*. When the compost is moist, the plant can often be gently teased out of its pot to examine the root system, which should remain in one piece.

Flowering

Different hybrids flower at different times of the year and will sometimes flower more than once. New flower spikes emerge from the leaf axils between the lower leaf and the pseudobulb.

A beautiful crystalline white Odontoglossum crispum ▶

Old backbulbs that have been removed during re-potting will often sprout a new shoot, and can be potted up and grown on to produce another plant.

Miltoniopsis phalaenopsis from Colombia is an attractive parent of many Miltonia hybrids ▶

29

Dendrobium phalaenopsis

With around 1,600 species, found mainly in eastern Asia and Australia, *Dendrobium* is the second largest orchid genus. With their characteristic tall pseudobulbs (sometimes referred to as 'canes') and alternate leaves, they are popularly known as bamboo orchids. The flowers have a characteristic 'chin', and the flower buds themselves are unusually attractive.

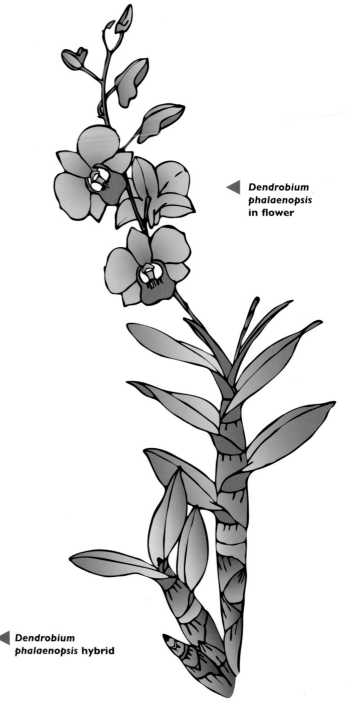

◄ **Dendrobium phalaenopsis in flower**

◄ **Dendrobium phalaenopsis hybrid**

A number of different types of *Dendrobium* are offered for sale as houseplants, the most common being *Dendrobium phalaenopsis*. They can be recognised by the long sprays of flowers emerging either from, or close to, the tops of the pseudobulbs. They are notable for their vivid colours, particularly a deep cerise or purple. White forms are also popular and they are becoming available in an increasing number of shades of salmon pink, lemon and green, often contrasting with dark, almost black, lips.

Dendrobium phalaenopsis prefers warm temperatures, 18 to 28°C (64 to 82°F), combined with bright light, and will flower well in these conditions. Direct sunlight should however be avoided in the summer to prevent scorching. As a general guide, they can be watered once a week throughout the year, and perhaps more in the summer. Ideally they should be re-potted in the spring in a similar manner to that described for Cambria types.

Dendrobium phalaenopsis are also sold as cut flowers by florists, and although they are now cultivated in many other countries in South-East Asia, they are sometimes sold under the name of Singapore orchids. Unlike those bred for cultivation in pots, where the aim is to produce a compact plant, cut flower *Dendrobium* have been bred to produce long sprays of blooms that last well in water. Orchid farms in Thailand for example may have many acres of *Dendrobium* growing under shade-cloth to protect the flowers from the tropical sun and produce perfect blooms.

Purple *Dendrobium* for sale in garden centre ▶

◀ Close up of a flower of a delicate pink-tinged *Dendrobium phalaenopsis* hybrid

◀ This *Dendrobium* farm in Thailand produces many thousands of cut flower spikes for sale each year

Lime green *Dendrobium* ▶

Paphiopedilum –
slipper orchids

Paphiopedilum are slipper orchids from Asia and New Guinea. They are known as slipper orchids because of their pouch-like lips. Although each new growth typically produces a single flower, there are varieties that bear multiple flowers. Some multi-flowered varieties such as *Paphiopedilum* 'Pinocchio' produce a succession of new buds as the flower spike continues to elongate. In this instance you should not cut the spike down to its base as soon as an individual bloom has died, but can allow the plant to produce several new blooms.

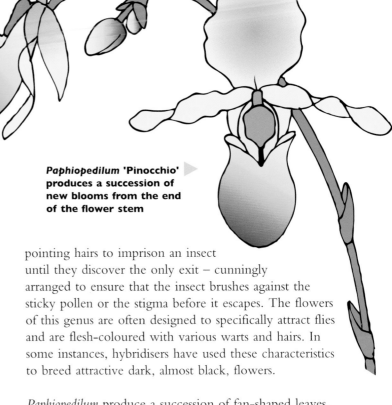

Paphiopedilum 'Pinocchio' ▶
produces a succession of
new blooms from the end
of the flower stem

Considered by many to be more bizarre than beautiful, people are often surprised to learn that *Paphiopedilum* are orchids at all. As 'trap flowers' they share many features in common with carnivorous pitcher plants that have also evolved to attract insects. The inside of the flower's lip is slippery with downward pointing hairs to imprison an insect until they discover the only exit – cunningly arranged to ensure that the insect brushes against the sticky pollen or the stigma before it escapes. The flowers of this genus are often designed to specifically attract flies and are flesh-coloured with various warts and hairs. In some instances, hybridisers have used these characteristics to breed attractive dark, almost black, flowers.

Paphiopedilum produce a succession of fan-shaped leaves, which are often attractively patterned. Because they don't have pseudobulbs, they are not able to store water in the same way as the Cambria types and appreciate a moist (but not wet) open but water-retentive compost. Nevertheless, many have evolved to survive dry conditions in limestone habitats, and their leaves are surprisingly tough, and to some degree succulent, with a thick waxy coat. They should not be over-watered.

▲ **Paphiopedilum niveum, a beautiful**
species from Thailand and Malaysia

Paphiopedilum normally grow on the forest floor in dappled shade and require less light than some of the other orchids offered for sale as houseplants. They generally prefer slightly cooler conditions and tend not to like centrally heated rooms. They can have a relatively small root system. They thrive in light, well-aerated compost and won't tolerate soggy conditions around their roots. Ideally they should be re-potted annually. Plenty of drainage material, such as gravel or polystyrene chips, should be placed in the bottom of the pot. Generally it is better not to give the plant too large a pot. Plants should be firmly planted, but on the other hand care should be taken not to compact the compost too much.

The flower spikes should be trained up a support as they develop to produce a nice straight stem. As the flower bud opens, place a further tie just behind the flower to lift it up, face forward and present itself at its best.

Orchids at risk

Paphiopedilum and *Cypripedium*: these slipper orchid cousins are particularly prone to illegal collection from the wild. The fabulous *Paphiopedilum rothschildianum*, which is only found on the slopes of Mount Kinabalu in Borneo, has been reduced to no more than a handful of plants in its remaining (secret) locations.

Paphiopedilum rothschildianum ▶

◀ **Paphiopedilum hybrid**

▲ *Paphiopedilum* grown in a
mixture of bark, charcoal,
Seramis® and perlite

◄ *Paphiopedilum*
hybrid

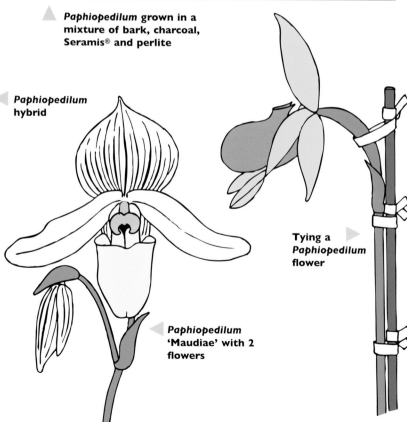

Tying a
Paphiopedilum
flower ▶

◄ *Paphiopedilum*
'Maudiae' with 2
flowers

Pleione

Originating in the cool moist habitats of the mountains of Asia, these delightful little orchids mainly bloom in the spring. They make ideal subjects for growing on the kitchen windowsill, and indeed are often sold as 'windowsill orchids'. A pot full of the delicate pale mauve blooms of *Pleione formosana* or one of its many hybrids is an ideal way to chase away those winter blues. Individual blooms normally last two or three weeks if kept in cool conditions, and the flowering season in your home can be prolonged for several weeks by selecting a range of different hybrids to grow.

▲ *Pleione pleionoides*

▲ *Pleione grandiflora*

Ready-potted pseudobulbs can be purchased in the spring, either in bud or in flower, or you can buy pseudobulbs of flowering size from late autumn until the end of February to pot up yourself, burying them so that one-third of the pseudobulb is still visible above the compost. Generally speaking, the bigger the pseudobulb the better. Larger pseudobulbs may have two or even three flower buds. Any dead roots should be trimmed to within half a centimetre (a quarter of an inch) of the bulb. Although it is sometimes recommended that dead bracts should also be removed, it may be wiser to leave them in place, as it is very easy to accidentally knock off developing flower buds.

Pleione are especially attractive when displayed in groups of three or five or more. They can be relatively shallow-rooted, and groups can be cultivated in half-height pots or 'pans'. The flower buds begin to develop before the roots and no water should be given until the flower buds begin to appear. Special care should be taken with watering in spring – you don't want the compost to be bone dry, but over-watering can soon lead to the death of the new roots. As *Pleione* only produce one flush of roots in the spring, this can be disastrous. Watering should be sufficient just to moisten the compost. Later in the season, as the plants build up a strong root system, watering can gradually be increased and feeding can begin.

Once flowering is finished, one or more leaf shoots develop, and gradually the pseudobulbs begin to shrivel as their reserves are used up. The new shoots will expand into large thin leaves. All feed should be applied at half the recommended strength. The addition of too much phosphate in the compost can lead to browning of the ends of the leaves. A balanced feed with an NPK ratio of 30:30:30 is suggested on a regular basis in the early part of the growing season. Towards the end of the summer this can be replaced with feed with a higher ratio of phosphorus and potassium as found in a tomato fertiliser. By September/October, when the pseudobulbs are plump, the leaves begin to show increasing signs of ageing and are eventually shed. Watering should be gradually reduced during this period and plants should be brought indoors into a cool, frost-free greenhouse or into the home before the first autumn frosts.

Pleione plants can be grown outside during the summer months. They can be taken out of doors after the last frosts (usually the middle of May) and placed in a shady location: too much sunlight can burn the thin leaves. If grown in the greenhouse, or in a cold frame, they should be placed under 50 per cent shade cloth, or its equivalent in glass whitening. Flower buds begin to grow towards the end of the summer so high temperatures are to be avoided, especially in July/August, otherwise the buds may abort.

Dormant pseudobulbs should be stored in a cool, dry place for at least six to eight weeks to get the full benefits of vernalisation (a treatment that will make them grow vigorously in the spring). A temperature of 5°C (40°F) or less (but not below freezing) is ideal, and can be achieved by placing the pseudobulbs in the garage or a shed, or in a sealed plastic bag in the bottom of the fridge.

▲ **Pleione limprichtii** growing in its natural habitat in Sichuan, China

A dormant Pleione bulb with two flower buds ▶

Potting *Pleione*

Pleione should be re-potted annually before they come into flower and any new roots begin to emerge. Use a compost that is open and free-draining but more water-retentive than a compost suitable for epiphytic orchids. A mixture of pine bark, sphagnum moss (for moisture retention), perlite, and perhaps a little oak or beech leaf mould, would make a suitable compost.

Pleione pests

False spider mites

So-called because they do not spin a web, the false spider mite *Brevipalpus oncidii* can cause considerable damage to *Pleione*. Scarcely visible to the naked eye, the grower may be unaware of their presence. They produce a toxic saliva, and the first symptoms of a problem may be that the plants fail to thrive, with pseudobulbs becoming progressively smaller with each passing year, producing fewer and fewer flowers. Although they can affect all plant parts, they prefer the undersides of the pseudobulbs of *Pleione*, where they lay their eggs in cracks and crevices.

By re-potting your *Pleione* in fresh compost each year, which is good horticultural practice, you have an opportunity to inspect the pseudobulbs for mites and their eggs. The pseudobulbs can be cleaned and, if necessary, treated with an insecticide.

Vine weevils

When grown out of doors, *Pleione* may also be prone to infestations of vine weevils. The grubs burrow into the pseudobulbs causing considerable damage.

▲ *Pleione forrestii*

Author's tip

Not all *Pleione* are spring flowering. *Pleione praecox*, with its delicately perfumed lilac blooms and attractive spotted pseudobulbs, flowers in the autumn after it has dropped its leaves.

1 Plant bulbs in half pans

2 Flowers appear before new roots. Take care not to over water at this stage

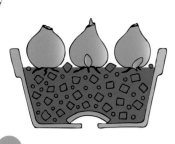

3 Water and fertilise freely in the growing period

4 Stop watering when the leaves have died and the bulbs have become dormant

Challenging orchids

Once you have mastered growing the easier types of orchids, you may like to have a go at something more challenging. There is an ever-increasing range of orchids for sale in garden centres and other outlets, but it is worthwhile seeking out specialist orchid nurseries too as they often offer a wide range of plants that otherwise you would not be able to find. The following are just a small selection of plants you might like to try.

Zygopetalum

Zygopetalum from South America are capable of filling a room with their heady perfume. Typically, a large ivory lip with purple markings contrasts with petals splashed with greens and browns. They have round pseudobulbs, prefer a well-drained compost, and need plenty of water throughout the year. Although only distantly related to *Miltonia* and *Oncidium,* in many ways they can be cared for in a similar manner.

Dendrobium nobile

Although in the same genus as *Dendrobium phalaenopsis*, the 'cool-growing' and deciduous *Dendrobium nobile* types are generally considered more difficult to grow in the home than their warmth-loving cousins. Frequently seen at spring orchid shows, they are easily recognised by their tall canes (pseudobulbs) covered with flowers. A well-grown specimen can have flowers along the full length of each mature cane, making a truly spectacular sight.

▲ *Dendrobium nobile* **hybrid spike**

▲ *Zygopetalum* **flowers are strongly perfumed**

43

As winter approaches, watering should be gradually reduced. The leaves will begin to yellow and eventually fall off. It is important that the plants are given a winter rest in a cool place to stimulate flowering the following year. If you water too soon, the young buds, instead of developing into flowers, will turn into small plants called keikis. Although sometimes seen on other orchids, keikis are most often produced along the flower spikes of *Dendrobium nobile* hybrids and flowering stems of *Phalaenopsis*.

A medicinal orchid

The accompanying photo shows *Dendrobium nobile* growing in China in a special area established exclusively for use as the source of the Chinese medicine called Shi Hu (which is the dried stems of *D. nobile*) used for stomach problems. The farmers nail the young stems onto local rocks called 'Dan Xia'. When large enough the stems are harvested, leaving the lower part of the plant and roots in place. The stems are then sold back to the local supplier, who dries them before selling them in the local markets. This is a truly sustainable sourced herb of a highly prized traditional Chinese medicine which also provides valuable income for local livelihoods.

▲ *Cattleya* **for sale in an orchid nursery in Brazil**

Cattleya

Originating in Central and South America, and named in honour of the English horticulturist William Cattley, *Cattleya* are the classic 'chocolate box' orchid with their large, perfumed and flamboyant blooms. Once popular for making corsages they have now been replaced by the more robust flowers of *Cymbidium*.

Once considered to be an orchid that could only be grown in greenhouses due to the large size of the plants, a new range of attractive hybrids suitable for growing on the windowsill have been produced in recent years. With their thick leathery leaves they enjoy high light levels, but full sunlight in the summer should be avoided to prevent leaf scorching. They can be grown at a temperature of up to 25°C (77°F) in summer but you should not let the temperature fall below 14°C (57°F) in winter.

Vanda

With their enormous showy blooms in shades of deep blue and violet, rose red or burnt orange, each with a tessellated pattern, *Vanda* are real show-stoppers. Definitely not for the beginner, *Vanda* seen for sale in garden centres may have been grown outside under shade cloth in Thailand for five or six years, before being transferred to Holland to be grown in greenhouses for a further year. Cultivated in baskets without any compost, their copious long thick silvery roots dangle in the atmosphere. They are often sold in large glass jars that help to maintain the necessary humid atmosphere around the roots. The advice is to fill the glass vase twice a week with luke-warm water until the roots are entirely covered, and leave for 30 minutes to give time for the roots to absorb the moisture. Orchid feed can be added to the water once a month.

Native to hot and humid climates, *Vanda* and their allies can be grown at temperatures between 17 to 28°C (62 to 82°F), and require high levels of light, but should not be placed in full sunlight in summer, as this will burn the foliage. These orchids are perhaps suitable subjects for a warm and humid conservatory rather than the windowsill.

This stunning blue *Vanda* hybrid, although not for the beginner, may be grown in a warm and humid conservatory

Masdevallia

With their charming triangular flowers resembling arrowheads, you may not recognise *Masdevallia* as being orchids at all. A wide range of hybrids is available, with colours ranging from hot, almost fluorescent, shades of red and orange to cool lemon-yellows and ivory. Originating in the cloud forests of Central and South America, they are definitely subjects for cultivation on a cool, shady windowsill on a bed of moist stones to provide the necessary high humidity.

Masdevallia often display bright, fluorescent colours ▶

Pests and diseases

Pests

Pest problems tend to be minimal within the home. Vigilance combined with prompt action should eliminate any problems before they get out of hand. If you can spot pests before they become a serious infestation, rather than resorting to pesticides, the culprits can be easily removed using a soft paintbrush, perhaps dipped in a weak detergent solution or methylated spirit. Many of us are understandably reluctant to use synthetic pesticides in the home due to their toxicity, and, in any case, they should always be used with caution. An alternative is to use a soft soap solution.

Mealybugs

Mealybugs are easily identified by their white waxy coats and the fact that they resemble miniature armadillos. Although they can be found on all parts of a plant, they are fond of hiding in the crevices of leaf axils or in the crowns. Indeed they are a particular menace in the crowns of *Phalaenopsis*, where only one or two individuals can soon cause considerable damage, which often leads to crown rot.

Mealybugs are easily removed with a soft paintbrush or a cotton swab dipped in methylated spirits or 70% isopropyl (rubbing) alcohol.

 Mealybug ▶

Aphids

Better known as greenfly or blackfly, aphids often appear as if out of nowhere. They are frequently found clustered on shoots or flower buds, or on the undersides of leaves. The first sign of an infestation may either be the appearance of discarded telltale white exoskeletons, which the aphids shed as they grow, or pale yellow spots appearing on leaves. They are sap-sucking insects and as they feed they secrete sweet sticky honeydew that encourages the growth of unsightly and damaging sooty moulds. Aphids seem to find *Cymbidium* buds and flowers particularly attractive and if untreated will soon leave unsightly brown blemishes on the petals. Heavy infestations can also lead to a distortion of the flowers. Minor infestations of aphids can be squashed or rubbed from shoots and buds using your fingers and thumb.

For heavier infestations you may choose to use an environmentally friendly insecticide such as a soft soap solution.

Red spider mites

Almost invisible to the naked eye, red spider mites are more easily seen with the aid of a magnifying glass. They are tiny members of the spider family that thrive in the warm, dry conditions of a conservatory or windowsill. The symptoms of red spider mite infestation begin with a typical silvery sheen on the undersides of leaves, which in serious infestations develops into a fine silk webbing. *Cymbidium* are particularly prone, and the undersides of their leaves should be examined regularly.

As an alternative to using an insecticide, red spider mites can be physically removed from the undersides of *Cymbidium* leaves by wiping them with a damp tissue. Grasp the base of the leaf firmly (young leaves can be pulled out of the plant all too easily) and wipe the leaf from the base towards the tip.

Scale insects

Resembling miniature limpets, adult scale insects are easily recognised. The eggs hatch beneath their protective umbrellas into tiny transparent 'crawlers'. Almost impossible to spot with the naked eye, they spread from one plant to another. The crawlers are probably the stage at which scale insects are most vulnerable to insecticides. The adults, with their protective coat, can only be dealt with by using a systemic insecticide. Prompt action and physical removal with a damp cloth are the keys to success. With patience and persistence, after a number of treatments, they are relatively easy to eliminate.

▼ **Coffee scale**

Slugs and Snails

Although not a problem in the home, slugs and snails top any list of pests in the garden. They have the ability to cause considerable damage to *Cymbidium* or *Pleione* that are grown out of doors in the summer months. Their position as the number one garden pest is reflected in the wide range of treatments available. The use of blue, toxic metaldehyde pellets is not to be recommended if plants are being brought into the home where young children and pets may encounter them.

Slugs like to curl up inside the bases of pots. A piece of mesh in the base will prevent them entering the compost and feeding on the roots.

Author's tip

Always carefully read the manufacturer's instructions on pesticides before use, and follow them carefully. Be aware that manufacturers may not test their products specifically for possible toxic effects on orchids.

Diseases

Although they are relatively uncommon in indoor plants, diseases can be encountered due to viruses, bacteria or fungi.

▼ **Fungal spotting on *Phalaenopsis* flowers can occur in warm, humid conditions where there is no air movement. This plant has spotting on the flowers probably because it has been on sale for a long time and the plastic sleeve has restricted air movement.**

Bacteria

Leaving water in the crown of *Phalaenopsis*, especially overnight when temperatures are cooler, can lead to bacterial infections and crown rot. This can be recognised by the centre of the plant turning brown and wet.

Bacterial infections usually cause leaves to become brown and spongy. Cut back the infected tissue to healthy green growth. Treat the cut surfaces with flowers of sulphur to prevent rot or further infection.

Viruses

The two most common orchid viruses are *Cymbidium* Mosaic Virus (CMV) and *Odontoglossum* Ring Spot Virus (ORSV). Viruses can cause colour breaks in flowers and/or sunken brown patches on the leaves, often in a characteristic diamond pattern. Unfortunately there is no cure for the above viruses, and the best advice is probably to dispose of the plant to prevent it contaminating others. Viruses are most easily spread from one plant to another by using contaminated tools.

Sterilise cutting tools such as secateurs or knives before use by soaking them in disinfectant or passing them through a flame.

Fungi

Phalaenopsis flowers are particularly prone to developing a grey spotting that indicates a fungal infection. This occurs in warm, humid conditions where there is a lack of air movement and can sometimes be seen where the plants offered for sale have remained in their plastic sleeves for some time.

Affected flowers should be removed and discarded, and the plants relocated to where there is more air movement.

What is an orchid?

Orchid flower structure

All flowers, including orchids, have the same basic structure. Each flower bud is protected by green sepals, which part to allow the petals to unfurl, revealing in their turn a circle of male stamens surrounding one or a number of female carpels (or pistils). At the tip of each stamen is an anther that contains the pollen grains. The carpels contain the ovules (eggs) that will, when fertilised by the pollen, become the seeds. At the tip of each carpel is a sticky platform or 'style' to capture the pollen.

Orchids have an outer whorl of three sepals and an inner whorl of three petals that together make up the flower. One of the three petals is often very showy and typically has been modified to act as both a flag and a landing platform for the pollinator. The sepals often resemble the other two petals in general shape and/or colour, and therefore the petals and sepals together are often called tepals.

Orchid flowers do not have distinct stamens and carpels as they have become fused to form a structure called the column. This combines both the male and the female sexual organs. In the majority of species, the pollen is found at the tip of the column in two, four, or sometimes eight, small lumps called pollinia, concealed behind what is known as an anther cap. The stigma is on the underside of the column behind the anther cap. When the pollinator (typically an insect) pushes its way past the anther cap into the throat of the flower the pollen becomes glued to its back. When it then visits another flower the pollen adheres to the sticky stigma.

Slipper orchids are a little different. The lip has become modified to become a distinctive pouch or slipper. The pollen is found in two sticky clumps, one either side of the column. The tip of the column is modified into a shield-like structure called the staminode and the stigma is located beneath the staminode. Two of the sepals are joined together to form what is known as a synsepal, which can be found behind the lip, while the third sepal arches over the lip.

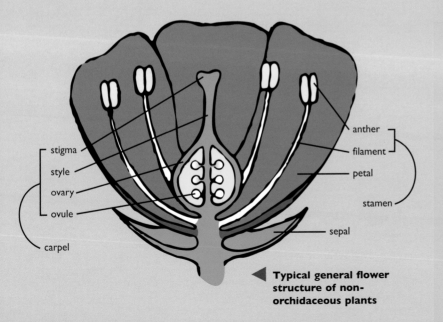

stigma
style
ovary
ovule
carpel

anther
filament
petal
stamen
sepal

Typical general flower structure of non-orchidaceous plants

Leafless orchids

Some orchids have sacrificed their leaves entirely, relying completely upon the chloroplasts in their fleshy roots for photosynthesis. These 'leafless orchids' include the elusive and ethereal ghost orchid, *Dendrophylax lindenii*, of Cuba and the Florida Everglades.

Fragrance

Although *Phalaenopsis* and *Dendrobium phalaenopsis* are not fragrant, and most *Cymbidium* hybrids exude only a slightly sweet aroma, there are some orchids that have very strong scents. Chief amongst these are *Zygopetalum* and some *Miltonia* and *Cattleya*, which can permeate an entire room with their heady perfumes at certain times of the day.

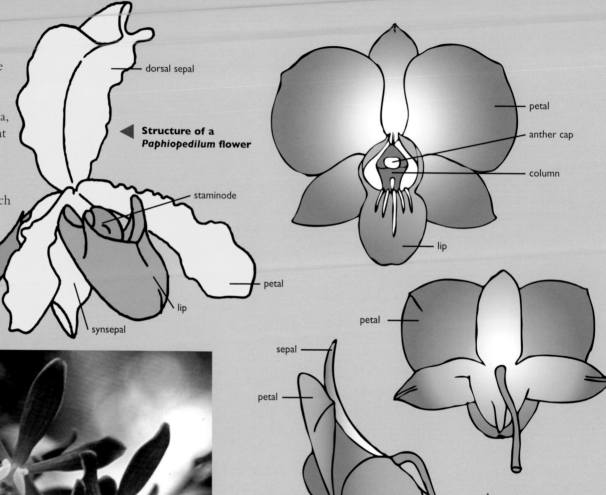

dorsal sepal

◀ **Structure of a *Paphiopedilum* flower**

staminode

petal

lip

synsepal

petal

anther cap

column

lip

petal

sepal

petal

lip

▲ ***Dendrobium* bloom showing the structure of a typical orchid flower**

◀ **Scent is often used by orchids to attract insect pollinators. *Encyclia phoenicea*, the pretty chocolate orchid native to Cuba, smells strongly of vanilla and chocolate in the early evening.**

Seeds

When it comes to seeds, orchids have followed the minimalist route. Individual seeds are minute, each seed weighing no more than a few micrograms. The contents of a seed capsule resemble talcum powder. What they lack in size however, they more than make up for in sheer numbers: a seed capsule may contain over a million seeds. Multiply the number of seeds by the number of capsules produced in a plant's lifetime, and a single plant can have an enormous reproductive potential. Charles Darwin calculated that within three generations, if all of the offspring of one orchid plant germinated and grew, it would 'clothe with one uniform green carpet the entire surface of the land throughout the globe.'

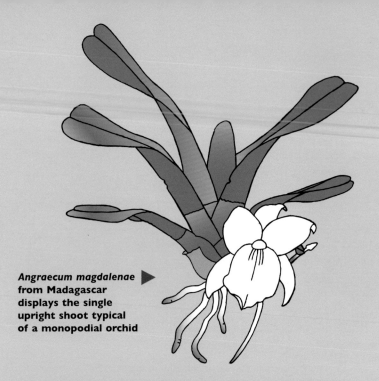

Angraecum magdalenae from Madagascar displays the single upright shoot typical of a monopodial orchid ▶

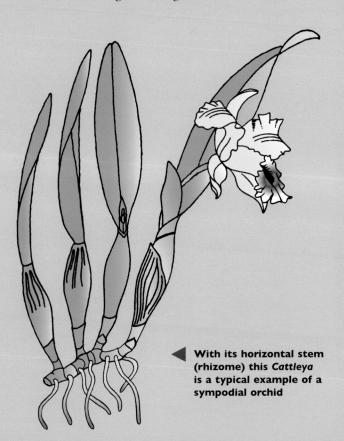

◀ **With its horizontal stem (rhizome) this *Cattleya* is a typical example of a sympodial orchid**

Leaves and pseudobulbs

Many tropical regions experience a distinct dry season rather than the cold winter familiar to those of us living in temperate regions. Because water may only be intermittently available, orchids often develop specially swollen stems called pseudobulbs that allow them to store water and survive dry periods. Other orchids, such as *Phalaenopsis*, have thick leathery leaves, which allow them to survive periods of drought. This means that orchids are often much more forgiving than other house plants regarding their needs for regular watering.

Growth patterns

Many orchids such as *Phalaenopsis* and *Vanda* have only a single growing point (technically they are described as being monopodial). Other orchids (including slipper orchids) are described as being sympodial – their growth habit is that of a horizontal branching stem called a rhizome, from which each successive shoot produces a new pseudobulb or fan of leaves.

53

Vanilla

The vanilla pods you can find for sale on supermarket shelves are the seedpods of the vanilla orchid. Native to Meso-America, the vanilla plant is actually a vine, which is unusual for an orchid. Different species of vanilla are now grown in tropical countries around the world, with some of the best quality beans reputedly derived from *Vanilla planifolia*, cultivated in large plantations in Madagascar.

▼ The Madagascan *Eulophiella roempleriana* with a flower spike that can be over a metre tall is definitely not an orchid for the windowsill

Epiphytic orchids

Most orchids are epiphytes, that is to say they grow on the trunks, branches and twigs of trees. Although known in the tropical regions of Central and South America as *parasitos*, epiphytic orchids only use the trees to give them a free lift up into the light, and are not parasites on the host plant. They often grow alongside a host of epiphytic lichens, mosses and ferns plus other flowering plants including, in the Americas, various bromeliads. Sometimes the sheer weight of epiphytes can cause a branch to break and come crashing to the forest floor.

The term 'rainforest' can be misleading as tropical forests range from seasonally dry forests, to cloud forests constantly dripping with moisture where epiphytes form aerial gardens that generate a thick layer of humus-rich soil on the branches, through to true rainforests with constant heavy rainfall all year round. Not only is there a wide range of different types of tropical forest, there are many different habitats within the forest. Orchids high in the canopy, such as many *Cattleya* species, are exposed through the daylight hours to the full effects of the merciless tropical sun. Under these arid conditions the plants have evolved adaptations similar to those displayed by cacti.

Cattleya quadricolor growing as an ▼ epiphyte in Colombia

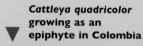

Rock-loving orchids

Lithophyte orchids are a small subset of epiphytic orchids adapted to growing on rocks, often in full sunlight.

Chinese slipper orchids

Sichuan in China is a hotspot for *Cypripedium* slipper orchids. Plants and hybrids raised from the seed of these orchids are beginning to appear for sale in specialist nurseries for cultivation either in cold greenhouses or out of doors, but do require specialist knowledge and care to grow them successfully.

▼ **A cultivated specimen of *Cypripedium macranthos* in China**

Terrestrial orchids

Terrestrial orchids grow on the ground, either in the shady conditions of the forest floor or in open grassland. Those commonly offered for sale include *Cymbidium* and tropical slipper orchids belonging to the genus *Paphiopedilum*.

Although the majority of orchids are found in the tropics there are also many beautiful species that grow in temperate regions including the British Isles. Britain has around 50 species of terrestrial orchid, some of which are every bit as exotic in appearance as their cousins in the tropics. Many grow in ancient meadows maintained by regional wildlife trusts. Temperate orchids such as *Dactylorhiza* and even *Cypripedium* are increasingly being offered for sale for cultivation in the cold greenhouse or outside in the garden.

◄ ***Paphiopedilum* plant**

Darwin's orchid

Charles Darwin was fascinated by the incredible variety of intricate pollination mechanisms of orchids, so much so that he wrote a book devoted entirely to the subject. He famously predicted the existence of a moth with a proboscis (tongue) sufficiently long enough to reach the drop of nectar in the tip of the 12-inch spur of *Angraecum sesquipedale*, the comet orchid of Madagascar. A night-flying hawk moth with a proboscis of this unprecedented length was discovered 21 years after Darwin's death.

A *Lycomormium* growing as a terrestrial in Ecuador ▶

◀ If you visit the Mediterranean in the spring, you can find many orchids in flower, such as this *Ophrys tenthredinifera* photographed in Mallorca in April

If you take the trouble to do a little research before you go on holiday, you can enjoy many orchids. The Mediterranean region is a hotspot for terrestrial orchid diversity, and attracts many orchid lovers in the spring when most species are in bloom. Although the plants and flowers may be small, those belonging to the genus *Ophrys* are particularly rewarding with their complex flowers that resemble various insects, including bees. These bizarre little flowers attract specific insects, which are duped into attempting to copulate with them (so-called 'pseudocopulation'), during which process the pollinia are glued to the unfortunate insect, which then flies off only to repeat the process on another flower, this time effecting pollination in the process.

Micropropagation

Growing orchids from seed

The dust-like seeds of orchids are dispersed on the merest breath of air. Food reserves in the seeds have been sacrificed in favour of miniaturisation. To germinate in their natural habitat the seeds need to encounter a suitable fungus, either in the soil in the case of terrestrial orchids, or on the bark of a tree if they are epiphytic. The chances of landing in a suitable environment and then encountering a compatible mycorrhizal fungus (mycorrhiza literally means *fungus root*) are small indeed, hence the need for an individual plant to produce vast numbers of seed.

If fortune favours a particular seed, fungal threads (hyphae) infect the embryo and effect germination. The embryo encloses the fungi inside special cells and digests the contents. The orchid remains a parasite of the fungus until it begins to make chlorophyll and is able to begin to photosynthesise and manufacture its own food.

Instead of the usual seed germination pathway of producing a root and then a shoot, the orchid seed swells and develops into a spherical ball of cells called a protocorm. If the orchid is an epiphyte, germination will easily take place in the light and the protocorm soon resembles a tiny green pea, eventually producing a shoot and its first roots. Many terrestrial orchids,

however, germinate below the soil surface in the dark and their protocorms are colourless. Green chlorophyll is not produced until the first shoot appears above the soil surface.

Orchid seeds therefore cannot be germinated in the normal way on compost. Fortunately, Lewis Knudson transformed the orchid world in 1924, when he developed a medium that would allow orchids to germinate without the aid of a fungal partner. He raised plants in aseptic conditions on a special jelly medium (called agar), which included all of the necessary nutrients for germination and seedling growth. Today most tropical species can easily be raised on one of a range of artificial media without the use of a fungus, and indeed it is possible to do this with simple equipment at home. Many orchids are produced by their thousands in laboratories by commercial nurseries in this way from seeds from selected parents.

Tissue culture

An alternative to raising orchids from seed is propagation in the laboratory by tissue culture. Each new shoot grows from a tiny ball of cells called a meristem. The meristem can be cut away from the plant under a microscope in sterile conditions and, in a similar procedure to that used with orchid seed, is transferred to a suitable growth medium. It can be divided repeatedly to produce a vast number of plantlets that are genetically identical to the parent (i.e. clones).

▲ **Transferring seedlings in a laboratory in China**

▲ **Micropropagated orchids raised in flasks**

Before the advent of this technique, clones of orchids, such as *Cymbidium,* could only be obtained by division. Because this was a very slow process such divisions were very expensive. Everything changed with the development of tissue culture techniques and many orchids are now produced in this way, including *Dendrobium phalaenopsis,* for the cut-flower industry.

Conservation

Sadly many orchid species are at risk of extinction in their natural habitats. Threats to orchid populations include the usual suspects of habitat destruction and land use change, together with the increasingly negative effects of global climate change. At the height of the Victorian 'orchid mania', collectors stripped thousands of orchids from the tropics before shipping them to Europe to die in overheated 'stove houses'. Happily, in some respects, we live in a more enlightened age but illegal collection still occurs particularly when new showy species are discovered. What is particularly distressing is that tropical orchids in general are easy to raise from seed and the collection of large numbers of plants from the wild is totally unnecessary.

Orchids are widely used, particularly in Asia, as traditional medicines and there is an increasing interest in cultivating these plants both to satisfy the expanding market and to reduce collection pressure on wild populations. Tubers of *Gastrodia elata* can be bought from street vendors in China, for example, where they have been used in herbal medicine for over 1,500 years. A remarkable plant, it has no chlorophyll and relies entirely on a soil fungus to obtain its nutrients from dead and decaying material. Chinese farmers plant the small tubers in fields sown with wood chippings, and harvest the new, larger tubers after two or three years.

There are many other positive initiatives around the world to conserve orchids. In terms of raising public awareness, many countries have adopted orchids as their national flowers; the beautiful *Guarianthe skinneri*, for example, is the national flower of Costa Rica.

One of the many successful orchid projects at the Royal Botanic Gardens, Kew has been the micropropagation and reintroduction of the British native *Cypripedium calceolus* into its natural habitat in the north of England. This elegant slipper orchid was reduced to just one individual plant in the wild due to over-collection. It was rescued by Kew's Sainsbury Orchid Project.

Kew is also involved in the Orchid Seed Stores for Sustainable Use (OSSSU) project, funded by the Darwin Initiative, which is establishing a global network of orchid seed banks with the aim of conserving the orchid floras of the world, particularly those of the biodiversity hotpots in the tropics.

Find out more about these projects and Kew's orchid collections by going to www.kew.org/orchids.

Important

Always remember, you should not dig up and remove orchid plants from the wild.

The Monteverde Cloudforest Reserve in Costa Rica ▶

▲ *Masdevallia rosea* **in Ecuador**

▲ *Laelia anceps* **in Mexico**

Importing orchids

When on holiday abroad you may be tempted to buy orchid plants to bring home with you. Within the European Union there are few restrictions regarding importation of orchids as it is effectively treated as a single country. Importation of plants into the EU (including the UK) is covered by CITES (Convention on International Trade in Endangered Species), and before even considering importing plants you should first consult their website (www.cites.org) as you will need to acquire the necessary import permits **before** you travel.

Orchids can only be imported without permits as cut flowers or when grown in flasks. However, if you want to import flasks of orchid seedlings you must have a Phytosanitary Certificate issued by the exporting country.

▲ *Cattleya maxima* **in Ecuador**

Cypripedium calceolus ▶

PLANTS PEOPLE
POSSIBILITIES

Photographers' credits: Christine Leon: wild *Dendrobium nobile* (and story) – page 44; Eric Young Orchid Foundation: *Dendrobium nobile* hybrid spike – page 43; Kanchit Thammasiri: outside back cover photograph of author; Chuck McCartney: Ghost orchid – page 54; Phillip Cribb: *Paphiopedilum rothschildianum* – page 36; vanilla pods, flower and harvesting page 54; Holger Perner: *Pleione* – photos on pages 39 and 40; Paul Little, Kew Photographer – cover, end papers, pages 6, 10, 11, 12, 13, 14, 15, 20, 24, 26, 30, 34, 36 – yellow *Paphiopedilum*, 46, 47 – mealy bug and aphid, 50.

First published January 2010

Re-printed March 2010

Royal Botanic Gardens, Kew
Richmond, Surrey, TW9 3AB, UK
www.kew.org

ISBN 978-1-84246-427-4

British Library Cataloguing in Publication Data.
A catalogue record for this book is available from the British Library.

Design and page layout by Culver Design

Printed in England by Beacon Press

For information or to purchase all Kew titles please visit
www.kewbooks.com or email publishing@kew.org

Kew's mission is to inspire and deliver science-based plant conservation worldwide, enhancing the quality of life.